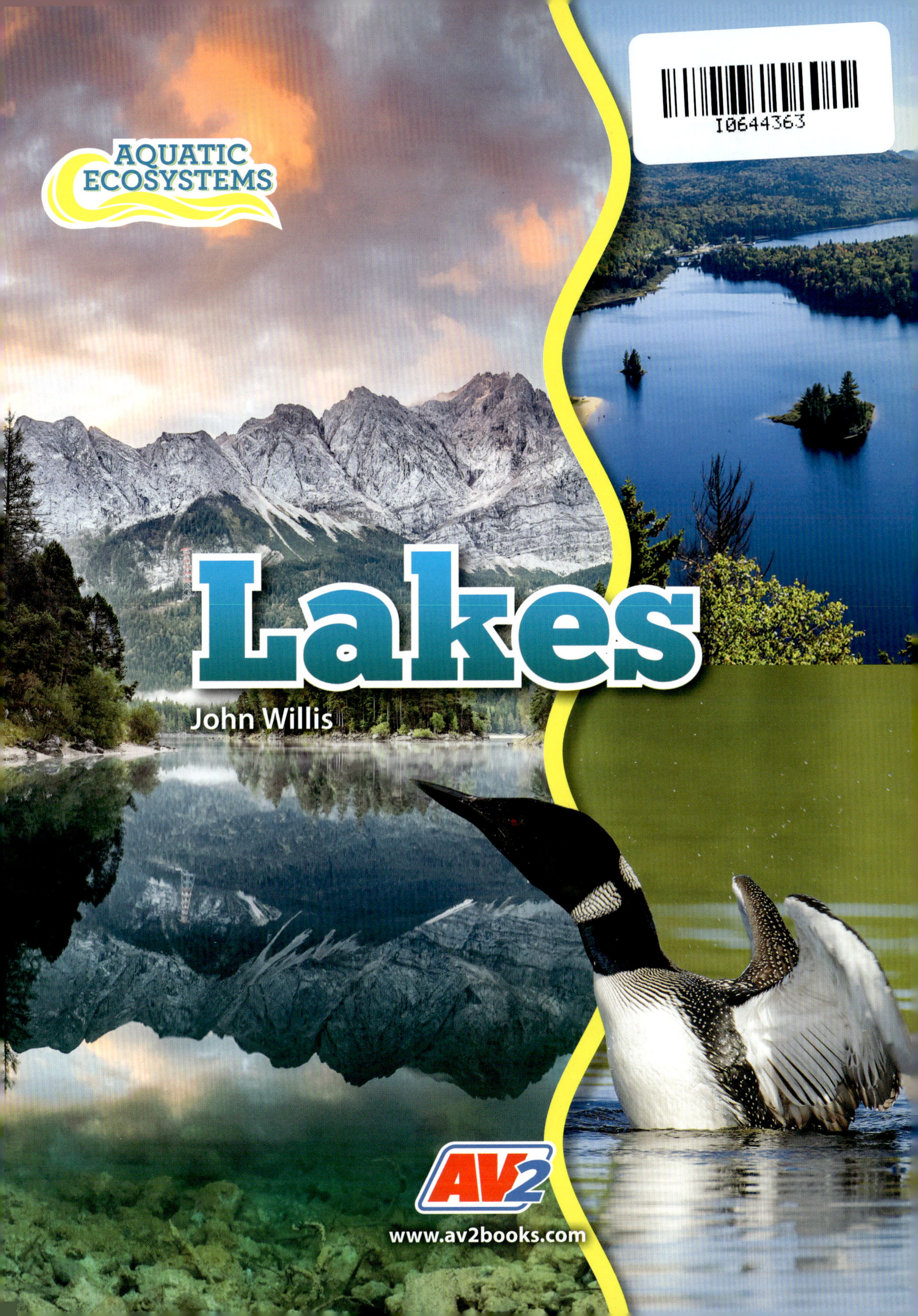

AQUATIC ECOSYSTEMS

Lakes

John Willis

AV2

www.av2books.com

Step 1
Go to **www.av2books.com**

Step 2
Enter this unique code

LIGEYHF7I

Step 3
Explore your interactive eBook!

AV2 is optimized for use on any device

Your interactive eBook comes with...

Contents
Browse a live contents page to easily navigate through resources

Audio
Listen to sections of the book read aloud

Videos
Watch informative video clips

Weblinks
Gain additional information for research

Try This!
Complete activities and hands-on experiments

Key Words
Study vocabulary, and complete a matching word activity

Quizzes
Test your knowledge

Slideshows
View images and captions

... and much, much more!

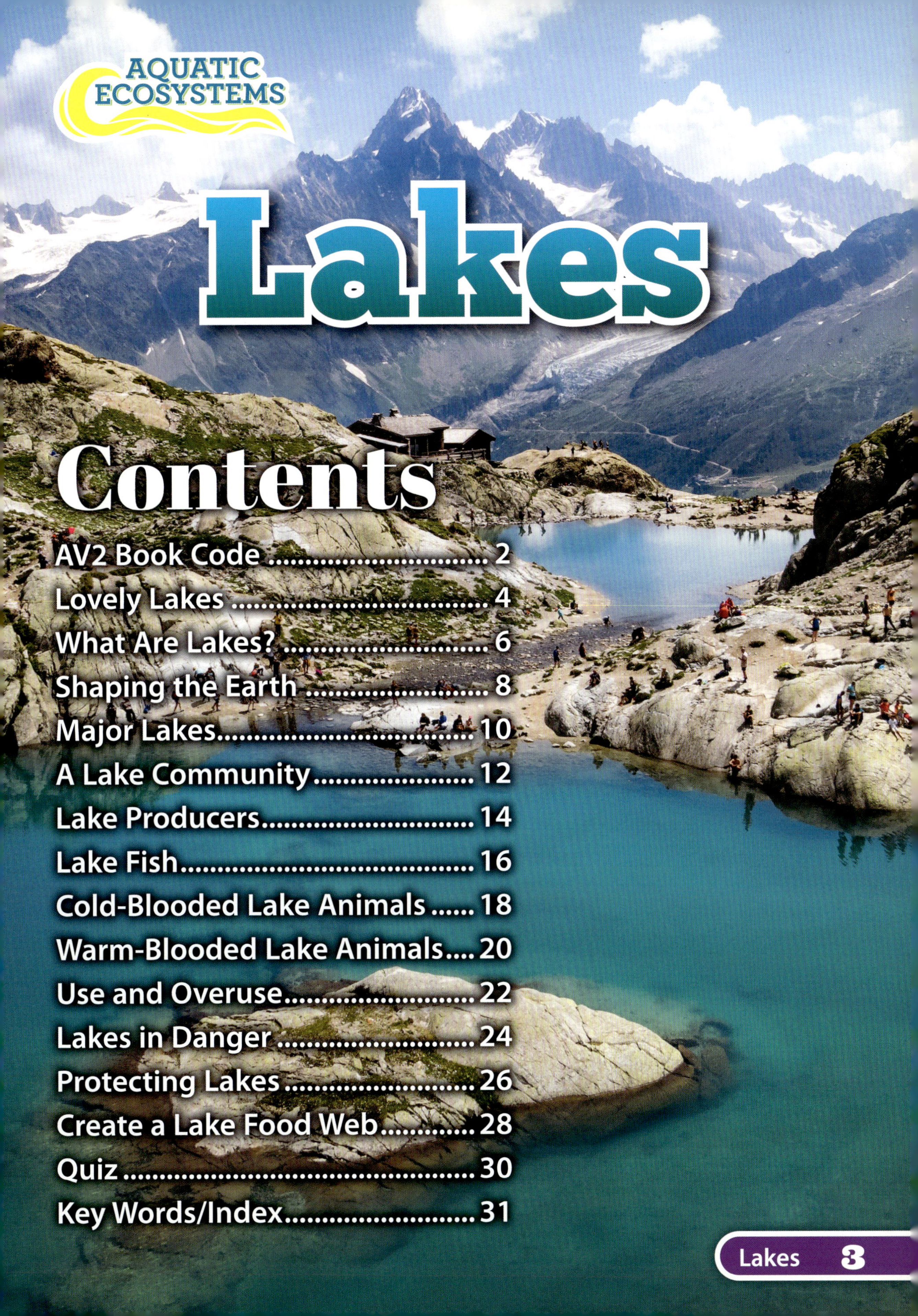

AQUATIC ECOSYSTEMS

Lakes

Contents

Lovely Lakes

Around the world, humans depend on Earth's millions of lakes. People rely on lakes for fresh drinking water, fish for food, and calm water on which to move goods and people. Larger lakes are often important shipping routes. People use lakes for relaxation, too. They tend to be popular sites for hiking, camping, swimming, and other leisure activities.

Lakes are also very important **ecosystems**. Many animals depend on the calm, still waters of lakes for food, water, and a home. While large numbers of lakes are freshwater lakes, others have salt water. Freshwater lakes hold less than 1 percent of Earth's total water, but they make up about 98 percent of the important surface waters that people can use. This is why balancing the needs of both humans and animals when using these ecosystems is an ongoing issue to this day.

Record-Setting Lakes

Highest Major Lake

Lake Titicaca, in Peru and Bolivia, is **12,500 feet** (3,810 meters) above sea level. It is the **highest** lake that is deep enough for large boats to travel on.

Deepest Lake

Russia's **Lake Baikal** is about **5,315 feet** (1,620 m) deep at its deepest point.

Saltiest Lake

Antarctica's **Don Juan Pond** is about **40 percent** salt. The oceans are only about 3.5 percent salt.

North America, Africa, and Asia have about 70 percent of the world's lake water.

What Are Lakes?

A lake is a body of water that is completely surrounded by land. Lakes come in many different sizes. Some lakes are small enough to fit in a room. They are often known as ponds. Other lakes are so massive that they can be seen from space. Many of these are known as seas. Although these different bodies of water have distinct names and sizes, they all share certain features with each other.

A lake collects water on Earth's surface in an area called a basin. Water collected in basins may come from **runoff** on Earth's surface or from underground **springs**.

The Dead Sea, between Israel and Jordan, is the lowest lake on Earth. It is about 1,300 feet (396 m) below sea level.

Many lakes' basins are made when glaciers move. These large, heavy pieces of ice leave large pits in the ground. When they melt, their water fills the pits. Other basins are formed by the movement of Earth's **tectonic plates**. When plates move away from each other, they can create large, steep basins. Volcanoes can also make lake basins. When a volcano erupts, it may leave behind a shape called a crater. When this fills with water, a new lake is formed.

Some lakes are known as open lakes. This means that water leaves them through rivers and streams, flowing elsewhere. Other lakes are closed lakes. There are no rivers or streams leaving them. Instead, water only leaves these lakes when it **evaporates**.

Crater Lake, in Oregon, is located in the crater left by a volcano that erupted 7,700 years ago.

Largest Lakes by Continent

Asia
Caspian Sea
149,200 square miles
(386,400 sq. km)

North America
Lake Superior
31,700 square miles
(82,100 sq. km)

Africa
Lake Victoria
26,600 square miles
(68,890 square kilometers)

Europe
Lake Ladoga
6,700 square miles
(17,350 sq. km)

Antarctica
Lake Vostok
5,400 square miles
(14,000 sq. km)

Australia
Lake Eyre
3,741 square miles
(9,690 sq. km)

South America
Lake Titicaca
3,232 square miles
(8,370 sq. km)

Shaping the Earth

Lakes can change the world around them in significant ways. One of the best-known ways is the effect that lakes have on temperatures in nearby regions. Water is able to take in, or absorb, heat very easily. It is also able to release heat well. This means water can act as both a **heat sink** and a heat source. Because of this, lakes can influence temperature greatly. In summer, large lakes, such as North America's Great Lakes, take in heat. During the winter, they release this heat into the air, warming the surrounding area. This is why areas around the Great Lakes have **moderate** climates.

Because the remains of animals sink to the bottom of lake basins, ancient lakes are good places to find fossils.

While lakes may seem to be permanent features of Earth's surface, they appear and fade as many years pass. A lake's basin will fill over time. When plants and **algae** in a lake die, they begin to break down, or decompose. Their remains sink down to the bottom of the lake. Sand, dirt, and small pieces of rock are washed into lakes by rivers or when it rains. Over time, lakes become smaller. Eventually, they turn into wetlands. After even more time passes, they become dry land.

The coastal areas around the Great Lakes are known as "The Fruit Belt." The effect the lakes have on the climate makes it easy to grow plants there.

Major Lakes

Lakes are found on every continent. They appear in both warm and cold environments. Many animals and plants depend on lake ecosystems for a place to live. Human settlements often appear near them as well. Today, some lake ecosystems are under threat from human activity.

Lake Michigan

Lake Michigan is the third largest of the Great Lakes and the only one that is entirely within the United States. Major cities, including Chicago and Milwaukee, are located near it. About 12 million people live on the lake's coast.

Loch Ness

Loch Ness has more fresh water than any other lake in Great Britain. The lake is 23 miles (37 km) long and is about 788 feet (240 m) deep. Many people believe that a monster, called the Loch Ness Monster or "Nessie," lives in its depths.

Lake Eyre

Lake Eyre is Australia's largest lake. Located in a desert, the lake is normally dry, holding only a small amount of salty water. Heavy rains can increase the level of water but are usually not enough to entirely fill it. On average, Lake Eyre is completely full about two times every 100 years.

Lake Victoria

One of the largest freshwater lakes on Earth, Lake Victoria has major problems with pollution. Factories near the lake dump chemicals into its waters, leading to the spread of harmful bacteria and making it more difficult for fish and other animals to survive.

A Lake Community

There are different lake ecosystems around the world. However, they all share some features. In every ecosystem, different **organisms** occupy certain roles. Energy passes between these plants and animals when they eat or are eaten.

Organisms called producers receive energy from the Sun. They turn this energy into a form that they can use. When other organisms eat producers, they take this energy for themselves. When an organism that eats a producer is eaten, the energy is passed on again. This means that all organisms in an ecosystem need producers for their energy.

Primary consumers are animals that eat producers. Most primary consumers are **herbivores**. **Carnivores** and **omnivores** feed on primary consumers. This is why they are known as secondary consumers. A tertiary consumer is a carnivore or omnivore that has no predators. Tertiary consumers get their energy from eating primary or secondary consumers. Finally, when organisms die, **decomposers** break down their remains. This recycles their energy. It turns into nutrients that producers can use.

A bluegill can grow up to 9 inches (23 centimeters) in length.

Lake Energy Pyramid

As energy passes through an ecosystem, some of it is lost. This is why there are fewer consumers than producers in an ecosystem. This relationship can be shown through an energy pyramid.

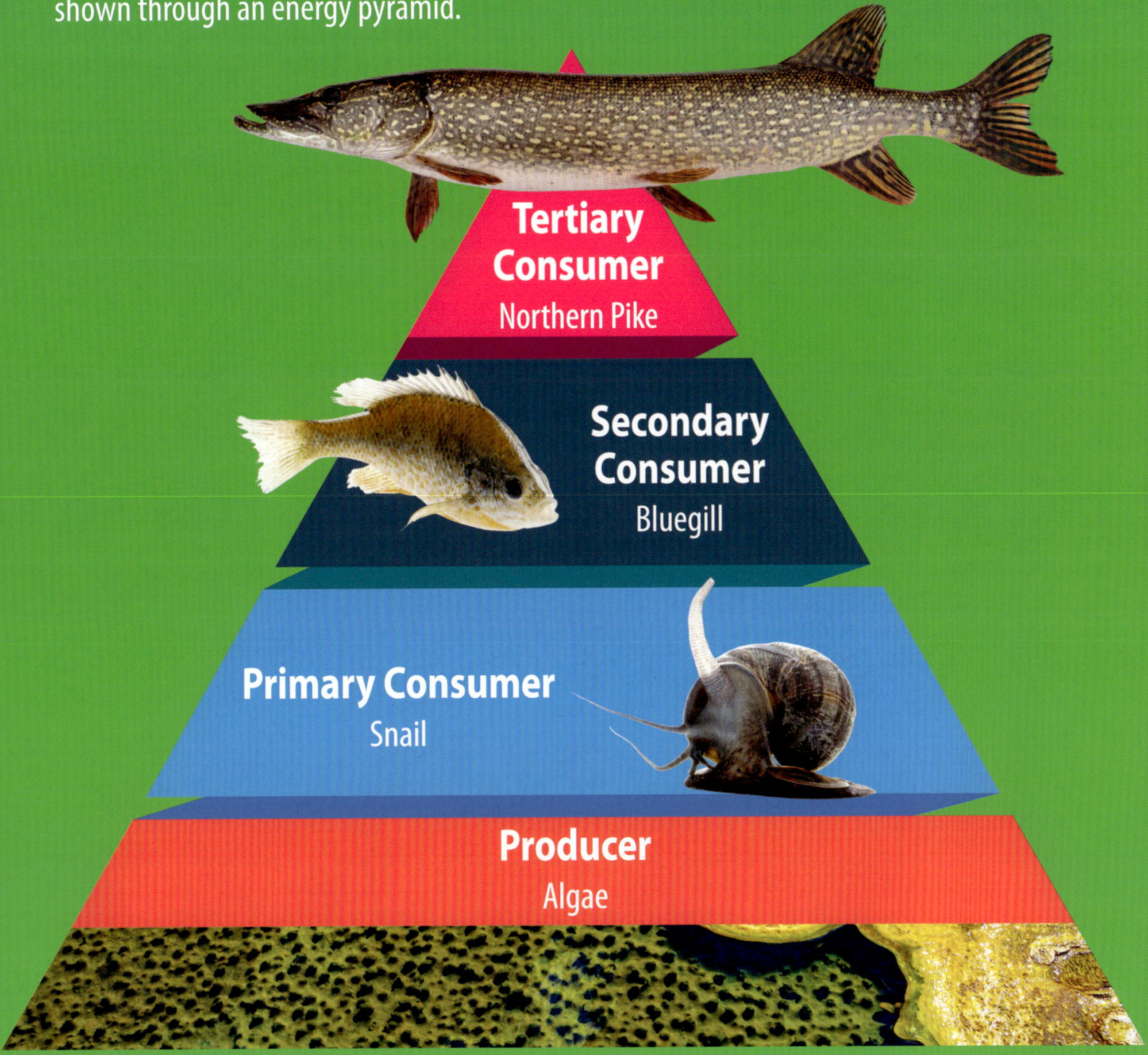

Try This!

Choose a lake plant or animal. Research it using the library or the internet. Is it a producer or a consumer? If it is a consumer, what kind is it? Use your research to create an energy pyramid that includes your chosen organism.

Lake Producers

By turning light from the Sun into food, plants and algae provide all the energy that other organisms in lake ecosystems need to live. Plants that live in an lakes and other aquatic ecosystems face challenges that plants living on land do not.

One challenge is finding sunlight. Deep under the water, there is little light. Plants often live in shallow parts of a lake in order to get the light they need. Because water in lakes is calmer than water in rivers or streams, many lake plants can safely float on the surface. This means there is no water blocking the Sun's light from reaching them.

In lakes with plenty of nutrients, hornworts form large groups that are homes for frogs, insects, and other animals.

Water lilies are a type of floating plant often found in lakes. They live in shallow water up to 8 feet (2.4 m) deep. A water lily's roots attach it to the bottom of a lake. Long stems grow up from the roots toward the water's surface. Attached to the end of these stems are large, wide leaves. Floating on the surface of the water, these leaves allow the plant to gather light for **photosynthesis**. Water lilies also serve as important homes for many animals by providing food and shelter.

Hornworts are another common type of lake plant. They are found all over the world. Unlike water lilies, hornworts live entirely underwater, where they float freely. While other plants use roots to get nutrients from the soil, hornwort does not have roots. It takes nutrients directly from the water. Hornworts can survive in cool waters with little sunlight. In places that freeze during the winter, they can survive underwater until spring.

The largest water lilies have leaves up to 6 feet (1.8 m) across.

Lake Fish

Different fish live in different areas of a lake. They find a **habitat** that suits their needs. Fish that eat plants or small insects can be found in littoral zones. These are shallower areas where there is enough sunlight for plants to grow. Other fish are found in open water, or the limnetic zone. Fish may also live at the bottom of a lake, known as the benthic zone. In deeper lakes, these fish may need to find food without any light.

Lake sturgeons are the biggest freshwater fish in North America. They live in the benthic zone of lakes. Lake sturgeons find food in the dark using whisker-like growths, called barbels, near their mouths. When sturgeons find food, such as clams or fish eggs, they suck it into their toothless mouths. Sturgeons have been on Earth for more than 200 million years. Lake sturgeons can grow to be more than 7 feet (2.1 m) long.

Many pike live in lakes as well. Unlike sturgeons, these fish are fast-moving predators that ambush their prey. Northern pike are found in North America, Europe, and Asia. They spend much of their time waiting among underwater plants or rocks, where their spotted skin helps them **camouflage**. When pike see another fish pass by, they lunge after it and catch it using their large, toothed mouths. Northern pike often eat fish, but will hunt any animal that can fit into their mouths.

Adult northern pike are about 20 inches (51 cm) in length.

Lake sturgeons can live for more than 150 years.

Some fish are only found in certain lakes. Lake Nyasa, also known as Lake Malawi, is one of the biggest lakes in Africa. The lake is home to more than 500 different **species** of cichlids. Many of these colorful fish live in this lake and nowhere else on Earth. Different cichlid species come in different shapes and sizes. They have adapted to live in distinct parts of the lake and eat various foods, including insects, algae, eggs, and even the scales of other fish.

Cold-Blooded Lake Animals

A cold-blooded animal is one that cannot maintain a constant body temperature on its own. Instead, it must rely on its surroundings to cool off or warm itself. These animals are also known as ectotherms. Many fish, along with reptiles, jellyfish, and arachnids, are ectotherms.

Lake Reptiles

Reptiles are ectotherms with scaly bodies and bony skeletons. They include lizards, snakes, and turtles. In order to warm up, many reptiles that live in lakes leave the water to bask in the Sun. By sitting in the light, they are able to warm up after the time they spent in the water.

Alligator snapping turtles are some of the largest freshwater turtles in the world. They live in slow-moving water, such as lakes, in the southeastern United States. Alligator snapping turtles spend most of their lives at the bottom of the water. They rise to the surface every 40 or 50 minutes to breathe. They catch food using their red, worm-shaped tongues as bait. When fish try to eat the bait, the turtle bites them using its powerful jaws.

Alligator snapping turtles can live for more than 70 years.

Kakaban Island is home to four different jellyfish species.

Lake Jellyfish

Jellyfish are soft-bodied animals that use stinging tentacles for defense and to catch food. Jellyfish usually live in saltwater lakes and oceans. Related animals, called freshwater jellyfish, are found in fresh water.

Kakaban Island, in Indonesia, is home to a large, **brackish** lake filled with jellyfish. Because these jellyfish live in a home with few predators, they no longer sting larger animals such as humans. Some swim upside-down. This allows algae in their bodies to use sunlight to make energy.

Lake Arachnids

Arachnids are small animals with eight legs. Some of the best-known arachnids on Earth are spiders. Many spiders use webs to catch food. One species, the diving bell spider, lives underwater and uses webs in a different way.

Each diving bell spider uses its web to capture a bubble of air and store it underwater. The spider lives in this bubble for most of its life. Diving bell spiders hunt small fish and other animals that come near their air bubbles. These spiders are found in plant-filled lakes around Europe.

Diving bell spiders carry an air bubble with them when they move underwater. The bubble's appearance is why their scientific name means "spinner of silver."

Warm-Blooded Lake Animals

Warm-blooded animals can maintain a constant body temperature. They are also known as endotherms. Both birds and mammals are endotherms. Many of these animals live in lakes or use them as an important source of food.

Lake Mammals

Seals are mammals that spend large amounts of time in the water. Almost all seals live in saltwater areas, such as the world's oceans. One species, the Baikal seal, lives only in the fresh waters of Russia's Lake Baikal. In winter, the lake's surface is frozen. Baikal seals use their strong claws to make holes in the ice. This allows them to reach the surface and breathe.

Some mammals, including beavers, muskrats, and otters, spend part of their lives in the water. These animals have thick, waterproof coats. This helps them keep warm in colder waters.

Baikal seals are small in size. They grow to a length of about 4.3 feet (1.3 m).

Land mammals may also rely on lakes for food. Moose are the largest members of the deer family. They need to eat large amounts of food, especially before winter, when they need to gain enough weight to survive the cold. One way they do this is by eating plants found at the bottom of lakes. A moose can dive more than 16 feet (5 m) underwater in search of food.

Lake Birds

Many birds can use their wings to fly. This allows them to change habitats, or **migrate**, when the seasons change. By doing this, they may have an easier time finding food or laying eggs. Birds that eat fish often rely on lakes for their food.

Loons are large diving birds that rely on water ecosystems, such as lakes, to survive. Their long, narrow bodies are designed to help them dive underwater, pushed by their strong, paddle-like feet. These birds also have heavy bones to help them stay underwater. Although strong swimmers, they do not walk well. Most of their time is spent in the water.

Bald eagles are large, powerful birds that mostly eat fish. Because of this diet, they often make their homes close to lakes. When these birds hunt, they use their long, wide wings to soar over a lake. Their powerful eyes allow them to see fish swimming under the water's surface. Bald eagles catch their food using the sharp claws, called talons, on their feet.

A loon can dive up to 200 feet (61 m) underwater to find food such as fish.

Use and Overuse

The St. Lawrence Seaway was completed in 1959.

Lakes have been important to people for many years and in many ways. In Canada and the United States, the Great Lakes are a major shipping route. The lakes are connected to each other, the St. Lawrence River, and the Atlantic Ocean by the St. Lawrence Seaway. This is a series of **locks**, canals, and channels. It allows large ships to travel from the ocean deep into the North American continent. Billions of tons of cargo have traveled through the Great Lakes since the Seaway first opened.

Lakes are also home to fish that make up an important food source for many people. Africa's Lake Victoria is responsible for hundreds of thousands of fishing jobs. The lake's fish have long helped to feed millions of people in Kenya, Uganda, and Tanzania, the three countries that border the lake. In more recent years, the number of fish in the lake has fallen. Today, people in the area must bring in fish from other countries for food.

Lakes have many recreational uses as well. People spend time swimming, boating, and fishing in the water. In the Great Lakes region of the United States, people spent about $16 billion on boating expenses in 2011. Each year, about 1.8 million people in the area fish for fun. While these recreational uses benefit people in many ways, they also affect the ecosystem of the lakes themselves.

Recreational Freshwater Fishers by Region

While people all around the United States enjoy fishing in fresh waters such as lakes and rivers, this activity is more popular in some regions than in others. Nearly 20 percent of freshwater fishers are from the East North Central region, which includes the Great Lakes.

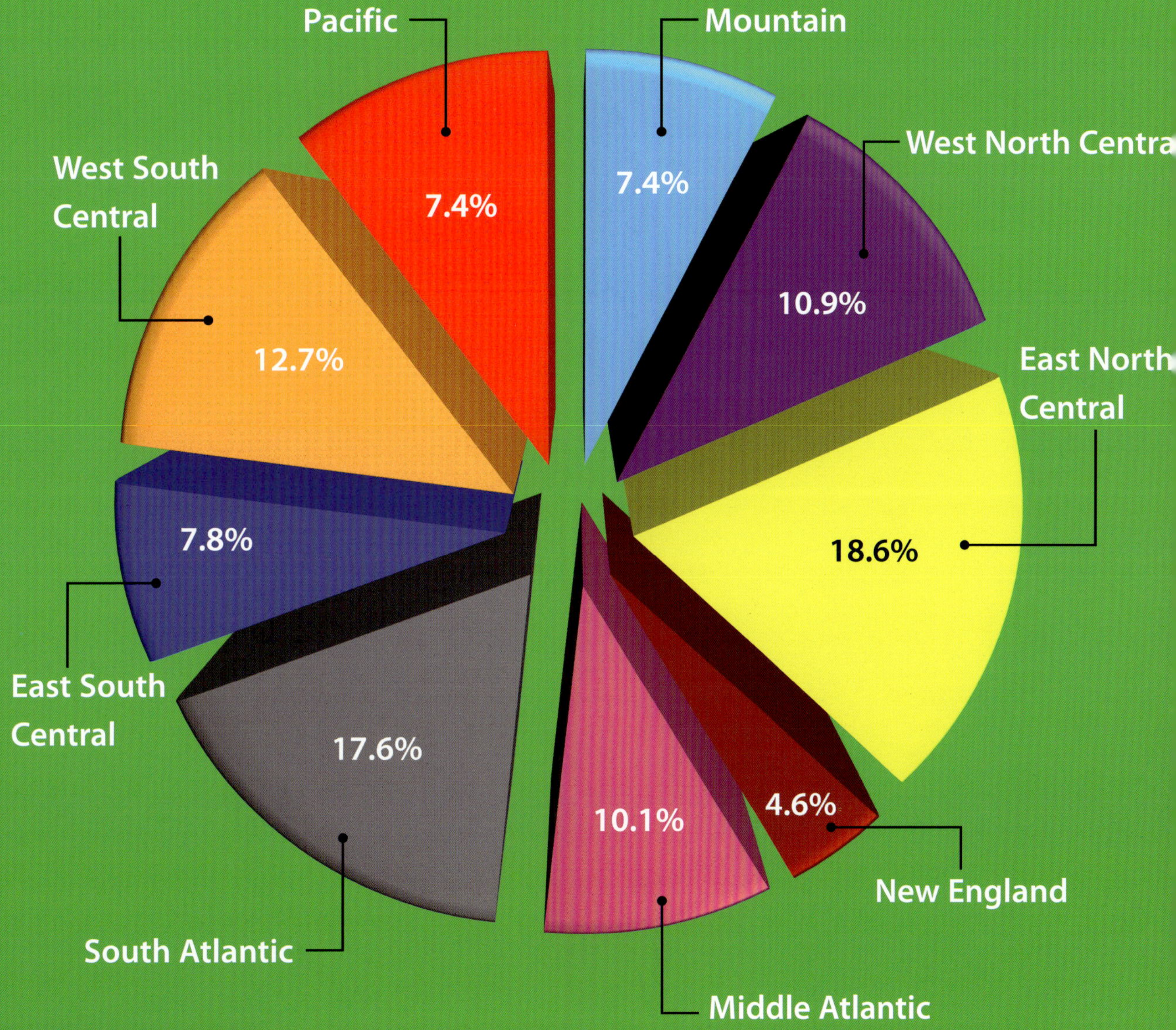

Think About It!

Many Americans enjoy fishing for fun and relaxation. What effect could this have on a lake ecosystem? What steps should people take to stop this activity from harming these ecosystems? Who should take these steps?

Lakes in Danger

When a lake ecosystem changes, the different plants and animals that rely on it may find themselves in danger. There are many ways in which these ecosystems may change.

Pollution in lakes can take many forms. For years, the area around Lake Karachay, in eastern Russia, was used to develop nuclear weapons. Waste from this production was dumped into the lake. The weapon plant was shut down about 30 years ago, but the place remains very **radioactive**. To this day, the lake is considered too dangerous for people to live near.

Some chemicals can help lake plants and algae grow too well. Phosphorous and nitrogen are used to help plants on farms grow. They are also dug up at mines. When these chemicals are washed into lakes by rivers or rain, they may cause a sudden growth of certain types of algae. This is known as an algal bloom. An algal bloom may block the Sun from reaching the lower parts of the lake. It may also alter the chemistry of the lake's water, making it hard for fish to live there.

Introduced species may compete with plants or animals already living in a lake for food. Often, these species do not have predators in their new ecosystems. During the 1800s, sea lampreys entered the Great Lakes area through human-made waterways. By the 1940s, they could be found throughout the Great Lakes. Because they have no predators of their own in this ecosystem, sea lampreys have greatly lowered the numbers of other fish throughout the Great Lakes region.

In their normal habitats, sea lampreys spend part of their lives in the ocean. The sea lampreys in the Great Lakes spend their entire lives in fresh water instead.

Algal blooms can make water poisonous to drink for land animals, including humans.

Protecting Lakes

There are many steps that people can take in order to help protect lake ecosystems. Some of these steps can be taken at home. People can help prevent algal blooms by making sure that chemicals from their homes do not end up in the water. This can include using only phosphorous-free fertilizers when planting gardens and cleaning up grass trimmings from their lawns.

Great Lakes Timeline

1930s

Annual fish harvests in the Great Lakes drop to less than 100 million pounds (45 million kilograms) due to large numbers of sea lampreys.

1954

The United States and Canada sign a treaty to form the Great Lakes Fishery Commission with the purpose of controlling the spread of sea lampreys.

1972

The United States and Canada commit to restore the Great Lake ecosystems and maintain safe drinking water by signing the Great Lakes Water Quality Agreement.

Protecting larger lakes can be much more difficult. Throughout their history, the Great Lakes have faced many challenges. These have included introduced species, pollution, and human development. Over time, many individuals and groups have taken steps to try to balance the need to protect this ecosystem with the need to use it.

The Great Lakes area has changed in many ways over recent years. How do you think the area will change in the future? What will cause these changes? Will these changes be positive or negative?

1985

With the Great Lakes Charter, eight U.S. states and two Canadian provinces agree to communicate with each other before using large amounts of water from the Great Lakes.

2008

The Great Lakes Compact, signed by the eight Great Lake states, protects plants, animals, and habitats throughout the Great Lakes area.

2019

The final design for the Fishpass project is revealed. This barrier is designed to let fish that live in the Great Lakes pass while stopping invasive fish.

Create a Lake Food Web

Plants and animals in a lake ecosystem rely on each other for food. This can be modeled using a food web. You can create a food web using the steps below.

1. Find at least three lake organisms of each type—producers, primary consumers, secondary consumers, and tertiary consumers.

2. Write down the name of one producer. Then, add the name of a primary consumer that eats it. Use an arrow to show the transfer of energy.

3. Write down the name of a secondary consumer that eats your primary consumer and the name of a tertiary consumer that eats your secondary consumer. Use arrows to show the flow of energy. This will create a food chain.

4. One at a time, add each of the remaining organisms to your food chain. Add new arrows to show which organisms they eat or are eaten by. Once all the organisms have been added, your food chain will be a food web.

Starter Food Web

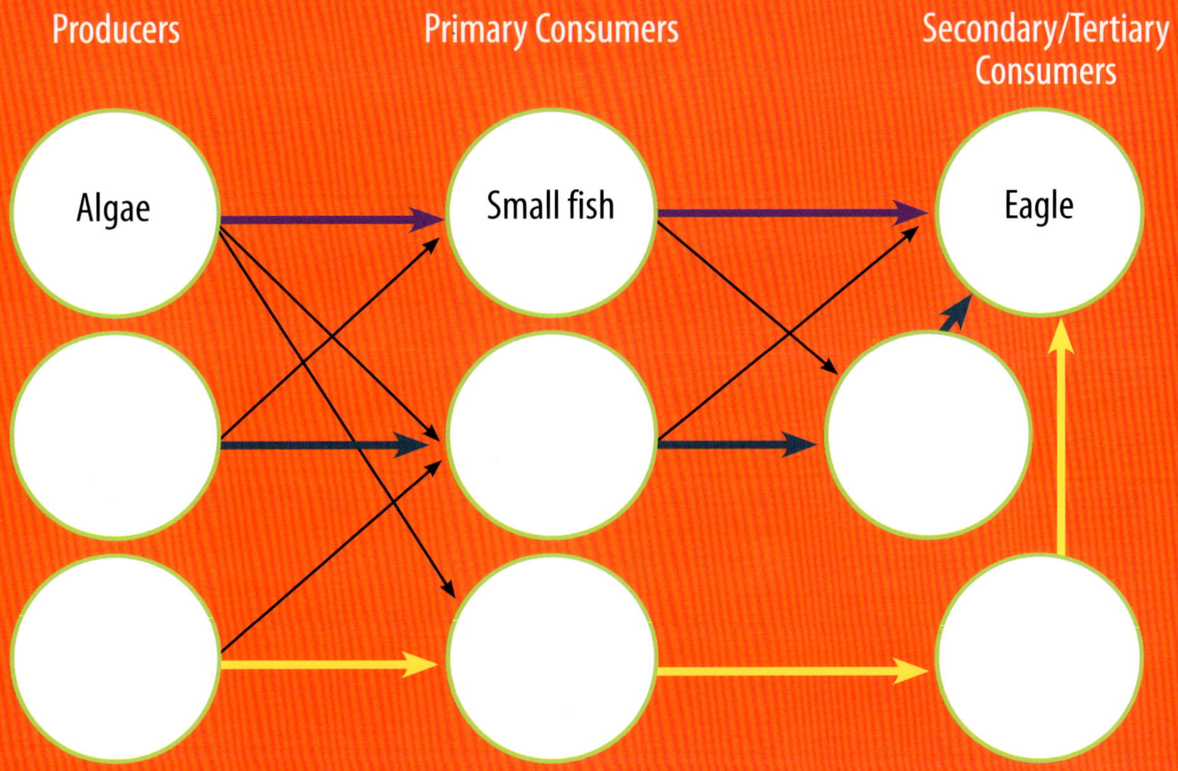

Once your food web is complete, use it to answer the following questions.

1. How would removing one organism from your food web affect the other organisms in the web?
2. What would happen to the rest of the food web if the producers were taken away?

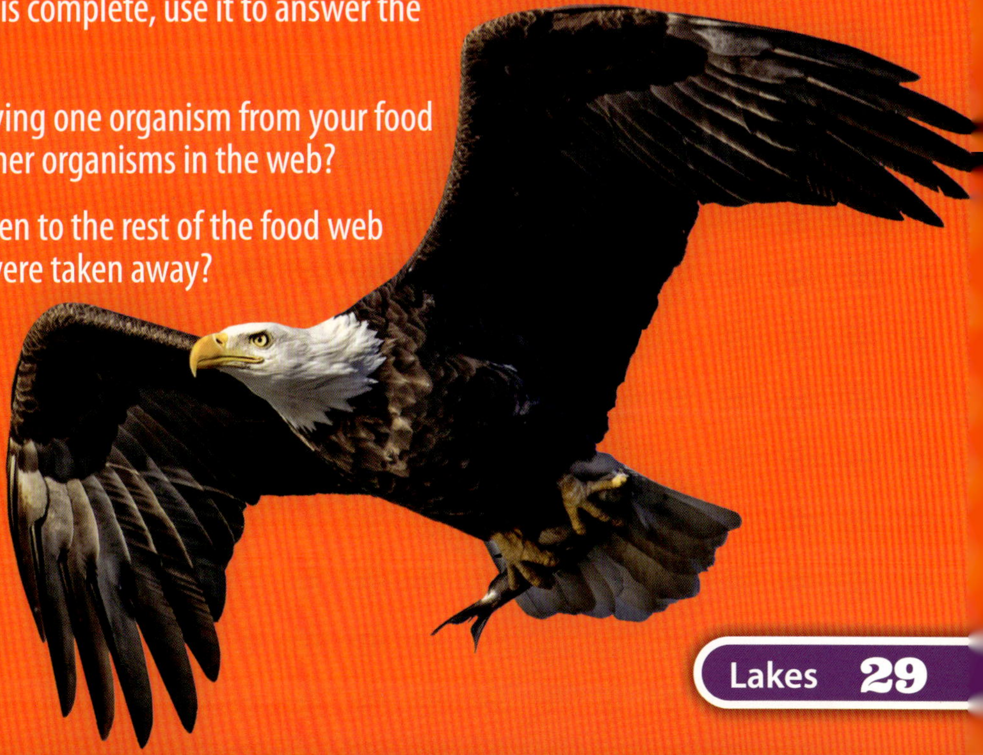

Quiz

1 How long have sturgeons lived on Earth?

2 What is the largest lake in Australia?

3 What is the name for a lake that has water leaving it through streams and rivers?

4 How many U.S. states signed the Great Lakes Charter?

5 How many countries border Lake Victoria?

6 What is the largest kind of deer?

7 What part of their bodies do alligator snapping turtles use as bait?

8 Which chemicals can cause algal blooms?

ANSWERS

1. More than 200 million years **2.** Lake Eyre **3.** An open lake **4.** Eight **5.** Three **6.** The moose **7.** Their tongues **8.** Phosphorous and nitrogen

Key Words

algae: plantlike organisms that can turn sunlight into energy

brackish: water with some salt but not as much as seawater

camouflage: when an animal hides in its surroundings

carnivores: animals that only eat meat

decomposers: organisms that break down dead matter

ecosystems: communities made of living things and the places in which they live

evaporates: turns from a liquid into a gas

habitat: the place where a plant or animal makes its home

heat sink: something that takes heat away from its surroundings

herbivores: animals that eat plants

locks: an area with gates at both ends that floods to let boats rise or lower between levels

migrate: when an animal travels to a new habitat

moderate: a climate that is neither too hot nor too cold

omnivores: animals that eat meat and plants

organisms: living things, such as plants and animals

photosynthesis: a process that produces energy using sunlight

radioactive: something that gives off energy in the form of radiation

runoff: water from the land draining into a body of water such as a river or ocean

species: a group of related animals that can have children

springs: underground sources of water

tectonic plates: large pieces of Earth's crust that make up the planet's surface

Index

Get the best of both worlds.

AV2 bridges the gap between print and digital.

The expandable resources toolbar enables quick access to content including **videos**, **audio**, **activities**, **weblinks**, **slideshows**, **quizzes**, and **key words**.

Animated videos make static images come alive.

Resource icons on each page help readers to further **explore key concepts**.

Published by AV2
14 Penn Plaza, 9th floor
New York, NY 10122
Website: www.av2books.com

Library of Congress Cataloging-in-Publication Data

Names: Willis, John, author.
Title: Lakes / John Willis.
Description: New York, NY : AV2, [2021] | Series: Aquatic ecosystems | Includes index. | Audience: Ages 8-14 | Audience: Grades 4-6
Identifiers: LCCN 2020017326 (print) | LCCN 2020017327 (ebook) | ISBN 9781791128159 (library binding) | ISBN 9781791128166 (paperback) | ISBN 9781791128173 | ISBN 9781791128180
Subjects: LCSH: Lake ecology--Juvenile literature. | Lakes--Juvenile literature.
Classification: LCC QH541.5.L3 W55 2021 (print) | LCC QH541.5.L3 (ebook) | DDC 577.63--dc23
LC record available at https://lccn.loc.gov/2020017326
LC ebook record available at https://lccn.loc.gov/2020017327

Printed in Guangzhou, China
1 2 3 4 5 6 7 8 9 0 24 23 22 21 20

052020
101319

Project Coordinator: John Willis
Designer: Ana María Vidal

AV2 acknowledges Getty Images, iStock, and Shutterstock as its primary image suppliers for this title.